Starfish

Summary
JOHN PATERSON

Giant Triton *(Charonia tritonis)*

The giant triton (*Charonia tritonis*) is a beautiful shell and a well-known predator of the crown-of-thorns starfish (*Acanthaster planci*). In many parts of the third world, it is still being collected in large numbers and sold to tourists as ornaments.

Many species of starfish are known to have outbreaks in different parts of the world. Prior to human collection, the giant triton might have controlled starfish numbers not by eating the many, but by preventing the aggregation that precedes the outbreak. At present, little is known of any aspect of the triton's ecology despite its obvious importance in controlling starfish numbers.

A PDF file of the complete book "Starfish: A warning from the past" can be downloaded from either Trove or the Internet Archive. Both hard and soft cover printed books are available through links at the Charonia Research blog.

When were crown-of-thorns starfish outbreaks first noticed?

It's 1957 and Japanese scientists travel to the island of Miyake-Jima to study a strange disease affecting a coral reef. The island's people have noticed a strange increase in the number of the large, venomous crown-of-thorns starfish. Over the following years they kill large numbers of the starfish attempting to protect the living coral upon which the starfish is feeding. The Japanese scientists noted that they had observed giant triton shells feeding on the crown-of-thorns starfish, in research published only in Japanese.

5 years later

This same species of starfish is noticed in increasing numbers at Green Island, off Cairns here in Australia. Over the next five years, the starfish outbreak will consume much of the living coral at Green Island and other reefs in the vicinity. The giant triton is again observed feeding on crown-of-thorns starfish and research in Queensland is commenced on the feeding rate and prey preference of the giant triton. Tests are done with three species of starfish; multiple specimens of each species are placed in cages with giant tritons.

The results showed that while each giant triton ate one crown-of-thorns starfish per week on average, it ate less crown-of-thorns starfish than another relatively common species of starfish. The research concluded that the crown-of-thorns starfish was not the preferred prey of the giant triton. This conclusion was further supported by similar research done overseas, despite the observation that giant tritons were often located on the Great Barrier Reef, and elsewhere, eating crown-of-thorns starfish.

30 years later

It becomes apparent that such a conclusion about prey preference of the giant triton is simply not justified, given the crown-of-thorns starfish often escapes complete predation because of its relatively high mobility. It is apparent that there is much confusion between prey capture and prey preference of the giant triton. It is suggested this distinction is relevant to control of prey.

40 years later

Australia is unsuccessful in its attempt to list the giant triton in Appendix 2 of the Convention in Trade in Endangered Species (CITES) because Japan objects on the grounds of 'no evidence' for the endangered status of giant triton.

50 years later
The Great Barrier Reef Marine Park Authority reports that outbreaks of crown-of-thorns starfish are a threat to the Great Barrier Reef and controls are needed.

60 years later
While the giant triton may be protected on the Great Barrier Reef and elsewhere in Queensland, there is still no evidence this protection has resulted in restored populations of the giant triton. However, there is evidence of continuing illegal collection and trade in Indonesia where it is also legally protected.

WARNING

BATHERS BEWARE OF
CROWN-OF-THORNS STARFISH.

STARFISH SPINES ARE VENOMOUS AND
CAN CAUSE SEVERE INJURY IF ANIMAL
IS STEPPED ON OR HANDLED.

PLEASE DO NOT DUMP OR BURY
STARFISH ON BEACH.

Dr Peter James (left) and Dr Robert Endean (right) at Green Island (early 1980s).

PRINCIPAL FINDINGS

Increased sub-tidal abundance of blue linckia (*Linckia laevigata*) could precede outbreaks of COTS.

similar larval development - Planktotrophic

same predator - Giant Triton

Linckia   Charonia   Acanthaster

On Heron and Wistari Reefs, most species of starfish did not occur below low water. Increased subtidal abundance of blue linckia (*Linckia laevigata*) could precede outbreaks of COTS (*Acanthaster planci*) due to many factors including similar larval development and the same predators.

#CharoniaResearch 1

Most species of starfish are rare, cryptic, toxic and in one case (*Acanthaster planci*) even venomous.

similar larval development - Planktotrophic

same predator - Giant Triton

Linckia   Charonia   Acanthaster

On Heron and Wistari Reefs, most species of starfish are rare, cryptic, toxic and in one case even venomous

These two protected reefs are not known to have experienced COTS outbreaks even though outbreaks have been recorded further south on the GBR.

#CharoniaResearch 2

The preferred prey is the species hunted and attacked preferentially by the predator.

similar larval development - Planktotrophic

same predator - Giant Triton

Linckia   Charonia   Acanthaster

The preferred prey is the species that is hunted and attacked preferentially by the predator. Mortality data alone can not indicate a predator's preferred prey species as predatory attacks can be unsuccessful and predation can also be sub-lethal

#CharoniaResearch 3

Heron Reef (23° 27′ S, 151° 57′ E) in Capricorn Group at southern end of the Great Barrier Reef.

similar larval development - Planktotrophic

same predator - Giant Triton

Linckia   Charonia   Acanthaster

Heron Reef (23° 27' S, 151° 57' E) lies in the Capricorn Group which is towards the southern end of the Great Barrier Reef. It is a lagoonal platform reef with a vegetated cay at its western end. The cay supports a tourist resort and research station. A harbour has been dredged in the reef.

#CharoniaResearch 4

The attack of the triton elicits an escape response by the starfish.

Starfish - Predator or Prey

similar larval development - Planktotrophic

same predator - Giant Triton

Linckia   Charonia   Acanthaster

The attack of the triton elicits an escape response by the starfish which, if successful, results in rapid prey dispersion with the loss of only a few arms. The escape response varies in its successfulness and is heavily dependent on (1) size and hunger of predator, (2) prey size and degree of cumulative prey injury and (3) physical composition and relief of substrate. #ChaioniaResearch 5

Species of coral reef starfish may trigger larval settlement in the giant triton.

Gastropod parasite on starfish

similar larval development - Planktotrophic

same predator - Giant Triton

Linckia   Charonia   Acanthaster

A number of species of coral reef starfish may trigger larval settlement in the triton. While previous studies managed to rear larva almost to the point of settlement, they could not produce settled larvae that crawled on the bottom. The cultured larva all died in the plankton and the missing link may well be another species of starfish.

#CharoniaResearch 6

"a complex twist to more typical asteroid life-history strategies." – Knott at al (2003)

similar larval development - Planktotrophic

same predator - Giant Triton

Linckia

Charonia

Acanthaster

"Within the asteroid family Ophidiasteridae, many species are capable of asexual reproduction as adults, particularly Linckia. The presence of larval cloning in species that also alternate between sexual and asexual reproduction as adults would be a complex twist to more typical asteroid life-history strategies."
Knott et al, Biological Bulletin (June 2003), 204: 246-255

#CharoniaResearch 7

Many eggs may never be fertilised when adult populations exist at low densities.

similar larval development - Planktotrophic

same predator - Giant Triton

Linckia

Charonia

Acanthaster

It is possible that many eggs are never fertilised when adult populations exist at low densities, such as at Heron Reef. On Heron Reef, and possibly the Great Barrier Reef in general, where many reefs exist in relatively close proximity, lecithotrophic genera such as Nardoa, Fromia and Echinaster appear to be more abundant than they are on atolls.

#CharoniaResearch 8

Just like the fertilization reaction in the sea-urchin.

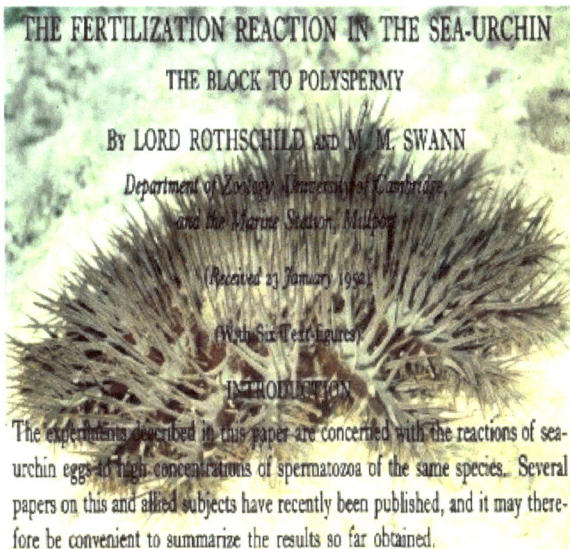

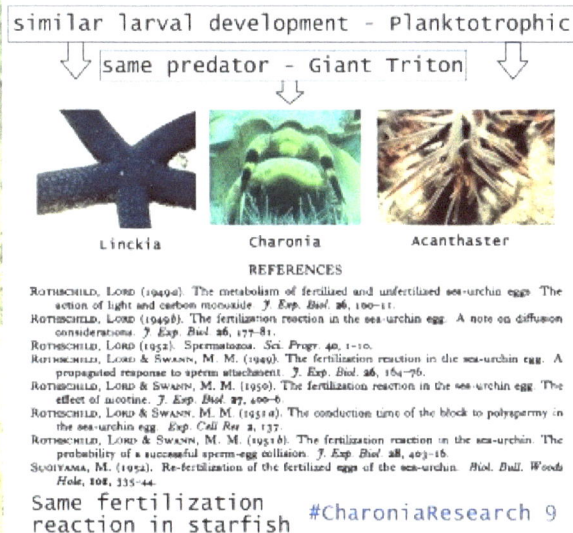

THE FERTILIZATION REACTION IN THE SEA-URCHIN

THE BLOCK TO POLYSPERMY

By LORD ROTHSCHILD and M. M. SWANN

similar larval development - Planktotrophic
same predator - Giant Triton

Linckia   Charonia   Acanthaster

Same fertilization reaction in starfish

#CharoniaResearch 9

SUMMARY OF RESEARCH

Sea-urchins and starfish both belong to Phylum Echinodermata and while early research on the fertilization reaction was conducted by Rothschild and Swann (1949) on sea-urchins, the conclusions regarding egg fertilization and proximity of spawning individuals were just as applicable to starfish.

Human collection of the Giant Triton and other predators was suggested by Endean (1969) as a causative factor in starfish outbreaks, but this Predator Control Hypothesis was generally disregarded due to the enormous potential numbers of starfish. Recent research demonstrating the strong avoidance reaction of the starfish to the triton together with an understanding of the importance of starfish aggregation to reproductive success may be slowly changing this opinion.

The existence of crown-of-thorns starfish outbreaks influenced many important world economic decisions of the 1960s, including the rejection on ecological grounds of a new sea-level Panama Canal. Starfish radial nerve extract (1-methyladenine) has been used to experimentally induce starfish spawning since Noumura and Kanatani (1962), but any possible causal connection with the starfish outbreaks has never been investigated.

Coral Reef Starfish and Giant Triton

Wistari Reef with Heron Reef in background

Heron Island with Wistari Reef in background

Acanthaster planci

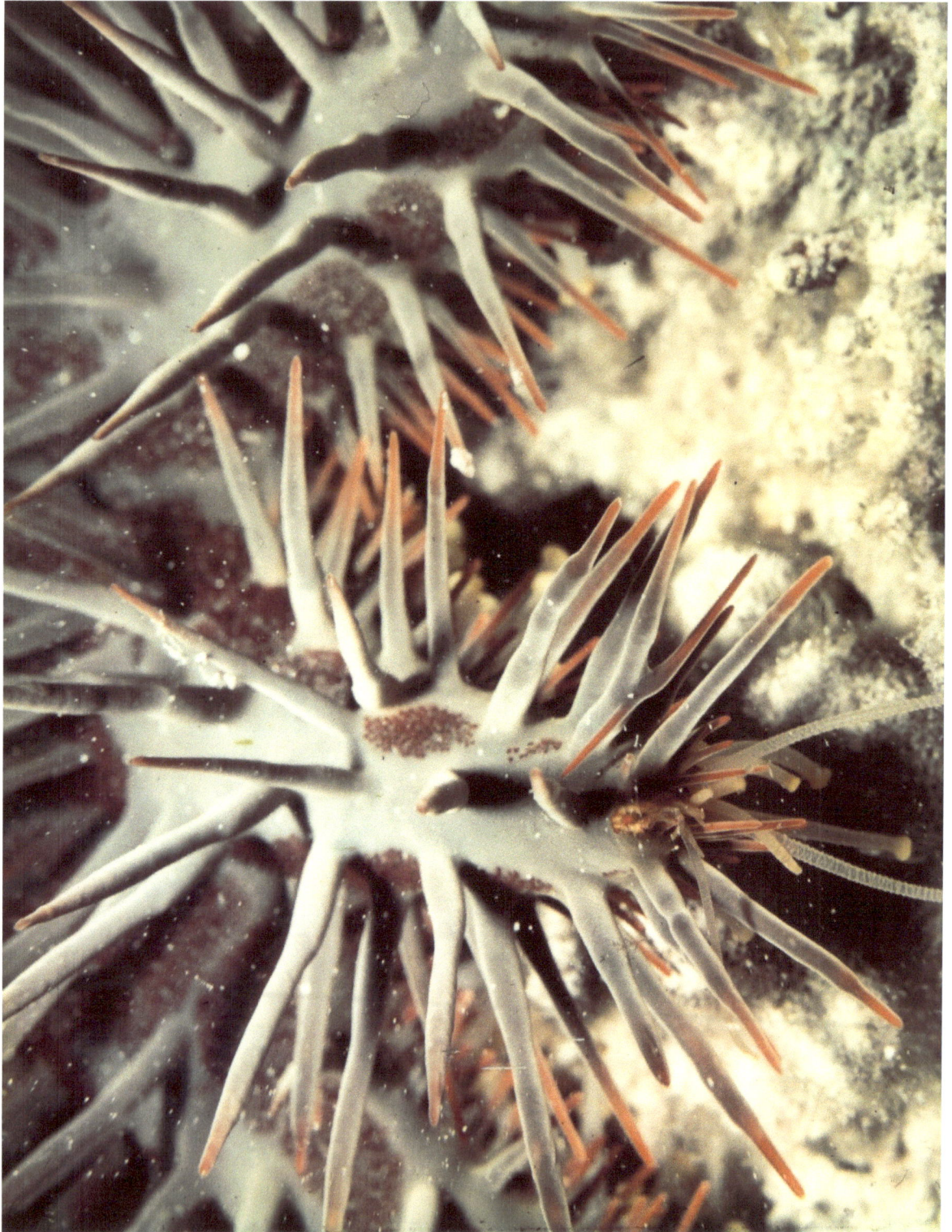

Acanthaster planci (Close up)

Fromia elegans

Euretaster insignis

Linckia multifora

Ophidiaster robillardi

Neoferdina cumingi

Echinaster luzonicus